Edgardo Roggero

Efficient Traffic Accident Reconstruction

Edgardo Roggero

Efficient Traffic Accident Reconstruction

Technological innovation for more accurate forensic investigation

ScienciaScripts

Cover image: www.ingimage.com

This book is a translation from the original published under ISBN 978-613-9-40851-1.

Publisher:
Sciencia Scripts
is a trademark of
Dodo Books Indian Ocean Ltd. and OmniScriptum S.R.L publishing group

120 High Road, East Finchley, London, N2 9ED, United Kingdom
Str. Armeneasca 28/1, office 1, Chisinau MD-2012, Republic of Moldova, Europe
Printed at: see last page
ISBN: 978-620-8-22039-6

Contents

To Sonia, my life partner and wonderful mother of our children.

PROLOGUE

The essence of this book goes back to my early days in a machine and structural design company, while simultaneously pursuing a degree in engineering. As I progressed through my studies, I discovered the exciting field of systems optimisation, a discipline I was introduced to by a lecturer who had played a leading role in the development of the iconic Concorde supersonic aircraft.

This discovery transformed my professional approach. I wondered how I had been able to design without knowing these advanced techniques, and this reflection radically changed my perspective: I went from creating simply feasible designs to developing truly optimised solutions. Optimisation captivated me from the outset, and I focused on discrete techniques, an area that, thanks to the growing power of computers, offered impressive new possibilities.

By combining these principles of optimisation with my experience in aerospace and forensic reconstruction, I have developed a methodology that transforms accident reconstruction. This methodology allows practitioners to accurately validate their hypotheses about the development of the accident, ensuring that the associated parameters are technically sound and indisputable. As a result, the presentation of evidence is strengthened, contributing to a fairer and more efficient justice system, with conclusions backed by solid mathematical foundations.

This book presents a rigorous and systematic methodology for accident reconstruction, offering a valuable tool for both professionals and those interested in the forensic analysis of road traffic accidents.

Throughout these pages, the reader will find not only a theoretical description, but also a case study to facilitate the understanding and application of the methodology presented. The primary objective is that this work will not only serve as a reference resource, but will become a constant companion for those who seek to unravel the truth behind every accident.

We hope that this effort, which integrates theory and practice, will serve as a guide for practitioners, students and anyone interested in the fascinating and crucial field of traffic accident reconstruction.

CHAPTER 1

INTRODUCTION

1.1 General

Traditional methods of accident reconstruction, despite their usefulness, have certain limitations. To overcome these limitations, this book proposes an innovative methodology, notable for its holistic approach, simplicity and accuracy. By combining the expert's experience with simple mathematical models, a realistic and scientifically based representation of the event is obtained without the need for complex software, making the accident reconstruction process more efficient in terms of the quality of the results versus the resources invested.

1.2 Definitions

Road Traffic Accident: A road traffic accident is an unforeseen event that normally occurs on public roads, involving at least one vehicle and resulting in property damage, injury or even loss of life. The aetiology of road accidents is multifactorial and can be attributed to human, vehicular and environmental factors, acting in isolation or in combination.

Traffic Accident Reconstruction: The reconstruction of a traffic accident is a scientific process that involves an exhaustive analysis of the physical and testimonial evidence to determine the sequence of events, identify the underlying causes of the accident and evaluate the responsibility of each party involved. Through the application of physical and mathematical principles, it seeks to objectively recreate what happened.

***Expert*:** In this text, the person responsible for the reconstruction of the road traffic accident will be referred to as the expert.

***Interested Party*:** the party requesting and financing the reconstruction of the accident to be carried out by the expert.

1.3 Main applications

The need to identify the causes of a road traffic accident can arise from both natural and legal persons, motivated by an interest in understanding the factors that actually contributed to its occurrence. They include authorities, judges, lawyers, public bodies, private companies and even individuals.

The main applications that motivate this need are listed below:

- Civil Liability Court Cases (when only economic damages are involved)
- Criminal Liability Court Cases (usually involving injury, or in some cases, death of persons)
- Improvement of Quality/Safety Standards (often required by automotive companies and/or related institutions).
- Simple Knowledge of Facts (typically required by private parties)

1.4 Reconstruction Results

When conducting an accident reconstruction, the Interested Party requesting the process generally expects to obtain the following information:

- What has been the contribution of each factor (human, vehicular and environmental) in the production of the event.
- Positions and speeds of the vehicles, as well as other possible players, at each key moment of the accident dynamics, especially at the beginning of the accident (see Figure 1.1 for an example).

Figure 1.1: Positions and Speeds during an Accident

Depending on their objectives, Stakeholders may also require (among others):

- A clear explanation of the origin of each damage, including the time of occurrence (before, during or after the accident). If the damage occurred during the accident, the specific time at which it occurred should be identified within its sequence of events (see Figure 1.2).

- Description of the driver's driving strategy.
- Under what conditions could the accident have been avoided or its consequences minimised?

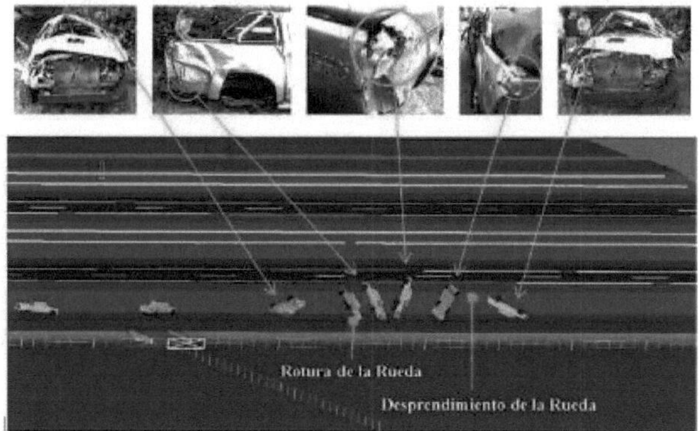

Figure 1.2: Correlation between Vehicle Positions and Vehicle Damage

1.5 Reconstruction Levels

The reconstruction of a road traffic accident will be more representative of the actual events the more in-depth and detailed the studies carried out. The quality of a reconstruction is therefore a function of three main parameters:

- Quality and volume of information available on the accident to be analysed.
- Skill, experience and dedication of the professional in charge of the reconstruction.
- Quality and quantity of resources/means used (including computational tools such as traffic accident reconstruction software or finite element programs, specific bibliographic references, among others).

Based on these considerations, four levels of reconstruction can be identified. The higher the level, the more accurate and reliable the match to the actual events:

Level 1: The reconstruction is based on

solely on the Expert's judgement and experience (without making any calculations).

Level 2: Based on simple calculations,

essentially using formulations based on the physico-mathematical principles of conservation of energy and quantity of motion.

Level 3: Complex calculations are used by analysing

step by step damage sequence; using specific accident reconstruction software (PC-CrashTM, Virtual CRASH among others) and also general applications applied to specific issues of the case under analysis (for example, programs based on finite element techniques, among others).

Level 4: Experimental verification of the case to be

is analysing (typically, it is based on repeating some event of the accident to validate a hypothesis, confirm some data and/or establish some conclusion).

Finally, it is important to note that the classification described above is closely linked to the cost and effort invested in the reconstruction. Although reconstruction has been divided into four levels, these are not watertight compartments; a specific case may require elements from more than one level, depending on its needs.

CHAPTER 2

Reconstruction Process

2.1 Synthetic Description of the Process

The reconstruction of a road traffic accident seeks to answer, in a precise and objective manner, three fundamental questions (through a rigorous technical and chronological analysis of the available information). These questions are:

a) (What happened?

b) (How did it happen?

c) (Why did it happen?

The higher the Reconstruction Level used, the more accurate, detailed and reliable the response will be.

Although the methodology described in this book addresses all three questions, the main objective of the book is to present a mathematically based approach to establish in a scientific way what happened and how the events unfolded during the accident, given its length, it does not address in detail how to determine the underlying causes of the accident.

Figure 2.1 presents a flowchart that summarises the recommended sequence and interrelation of steps for the efficient reconstruction of a Level 2 road traffic accident. This level has been chosen because the quality of its conclusions is sufficient to solve the vast majority of practical cases, and because it maximises the ratio: *quality of the results / resources invested.* It should be noted that the results obtained would not be altered if higher levels of reconstruction were used, only complementary information and/or a greater degree of detail would be available.

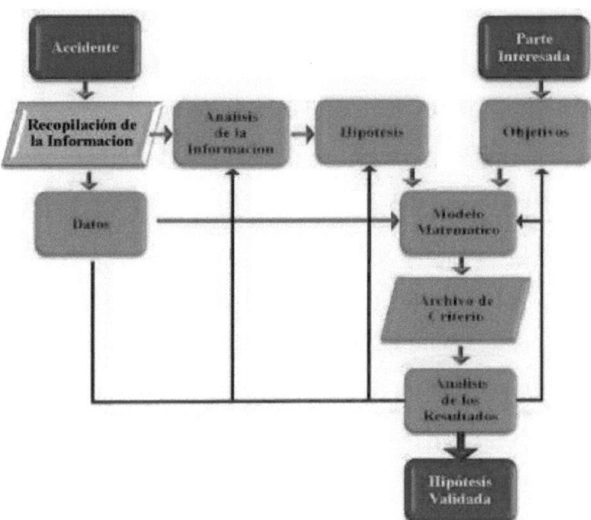

Figure 2.1: Reconstruction process

As can be seen in the figure, after the accident, the expert must collect all accessible information as quickly as possible in order to avoid loss or degradation of the information. Subsequently, he must analyse it in order to be able to formulate a hypothesis, which is able to explain how the accident occurred.

In the upper part of the figure there is a box called Interested Party, which represents the applicant of the reconstruction (Judge, Lawyer, Company, Institution, etc.) and consequently is the one who defines and/or proposes the objectives of the reconstruction. These describe what he/she wants to know specifically, for example: velocities, trajectories, instant of the dynamics in which damage occurs, etc. During the formulation of the hypothesis, the expert - conditioned by the person who requires his services - must consider these objectives, which are usually associated with the questions to be answered in the legal file in which the accident is being processed.

In order to validate the hypothesis and achieve the stated objectives, the expert must create a mathematical model, which must be fed with the available data in order to obtain the values of each parameter of interest.

This mathematical model, when solved, provides a set of feasible solutions that are presented in order in the "Criteria File". These feasible solutions are ordered from the supposedly best to the supposedly worst, according to the criteria established by the expert. These solutions are analysed and if they are completely consistent with the available information, the hypothesis will be accepted as the most likely explanation of the facts. Otherwise, it must be discarded or at least reformulated, restarting a new cycle from the most appropriate point (for example, requiring additional or more precise data; analysing the information from another approach, redefining the parameters of interest, the hypotheses, the objectives and/or the mathematical model. This iterative process concludes by identifying the hypothesis that explains all the events of the accident, and is consistent with all the available information, it can then be stated that this hypothesis has been validated.

2.2 Compilation of Information

2.2.1 Organisation of Information

For a better organisation of the information during the reconstruction of a road accident, it is convenient to use a Haddon Matrix specifically designed for this purpose. This matrix is composed of rows representing the factors involved, and columns indicating the points in time at which these factors should be analysed. Being the three fundamental factors: Human, Vehicle and Environmental, these factors should be analysed in three specific periods: Before the Accident, During the Accident and After the Accident.

		EVALUATION PERIOD		
		Pre-Accident	During the Accident	Post-Accident
FACTOR	Human	A	B	C
	Vehicular	D	E	F
	Environmental	G	H	I

Table 2.1: Haddon Matrix

The association between the above-mentioned factors and periods gives rise to nine different combinations, which are presented coloured in Table 1.1. The darker the shade, the greater its relevance in the reconstruction process, which implies that a greater effort has to be made to identify precisely and in detail the corresponding information.

The following is an example of the type of information that should be collected in each box:

A. Driver(s) driving experience, intelligence, health, fatigue, driving time, alcohol consumption, traffic violations, driving strategy, etc.

B. Ability to perceive potential hazards, ability to define avoidance tactics, how he/she acted in particular during the accident, etc.

C. Assessment of the state of the driver, witnesses and passengers after the accident in order to evaluate their testimonies, injuries, alcohol level, etc.

D. Condition of the vehicle prior to the accident, age, lack of maintenance, worn tyres, bad lights, mechanical damage, etc.

E. Damage caused during the accident, both mechanical and bodywork damage, scratches/marks, grazing, fluid loss, etc.

F. Evaluation of the condition of the vehicle after the accident in order to have objective information, state of conservation of the vehicle and the state of the vehicle.

after the event, etc.

G. Road/road condition before the accident, signalling, visibility, lighting, lighting, traffic lights, traffic lights, traffic, potholes, etc.

H. Atmospheric conditions, road conditions, signalling, traffic, objects on the road, roadside conditions, slopes, etc.

I. Assessment of changes in the environment in order to make them compatible with those existing at the time of the accident, etc.

2.2.2 Sources of Information

To complete this matrix, sources of information are available and can be classified into three categories according to their origin; External (information specific to the case under analysis that is not generated by the reconstructionist); Internal (that which the expert collects personally) and General (that which comes from relevant bibliographic references, both specialised and generic).

- External Sources

These include:

- Sketches/expertise of the Competent Authority
- Sketches/expertise/reports of other Experts
- Photographs and/or Videos of the Accident (locations,

status and positions of vehicles, etc.)

- Statements of Persons Involved (drivers, passengers, witnesses)

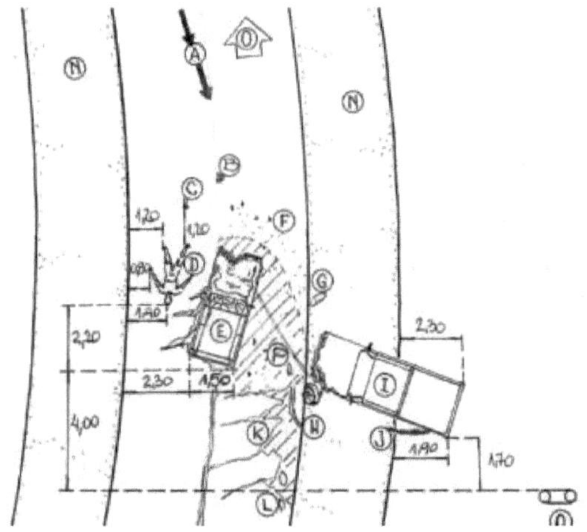

A. Arrow indicating direction of travel and trajectory of the vehicle

B. Remainder of the polarisation film.

C. Headrest.

D. Lifeless body of Selin Pablo Salma.

E. Automobile make: Toyota, model: Etios

F. Rest of acrylics y trisée glasses.

G. Rest of acrylics.

H. Rest of van bumper.

I. Pickup truck make: Fiat, model: Toro

J. Left rear wheel track of the pick-up truck.

K. Green striped area, representing s/oil stain.

L. Other opticians.

M. Arrows indicating direction of rotation of the van

N. Benches .

O. Arrows indicating the direction of circulation of the artery.

P. Print of effractions, printed on the lower parts of the van.

Q. Double cement poles of the electricity line with inscription C.

Figure 2.2 - Example of a Police Sketch with References

The competent authority sketches/experiments are key to the definition of trajectories, speeds and possible perpetrators, but they condition and/or complicate the performance of a proper analysis, in particular if they have one or more of the following characteristics:

✓ Lack of precision

✓ Lack of scales

✓ Absence of important measures and/or data

✓ Incorrect and/or contradictory information

Despite these limitations, the police sketch has the usefulness of serving as a basis (when complemented by other evidence such as photographs, statements, etc.) for the construction of hypotheses that confirm or refute the versions of those involved and/or witnesses with a view to identifying the primary characteristics of the accident on the basis of consistent and verifiable information.

<u>Internal Sources</u>

These include information obtained through:

- Ocular inspection Vehicles
- The visual inspection of the Accident Site
- Other items of interest

Unless there are severe limitations, the expert will visit the accident site to make a forensic analysis of the environment. He/she will also - if possible - inspect the vehicles to identify the damage suffered by the units. It is in these activities that the expert needs to act efficiently, since both the environment and the vehicles may have changed or possibly been tampered with since the time of the accident. These inspections also make it possible to validate and/or highlight errors or omissions in the information obtained from external sources.

The contrast between the following photographs is evidence of the modifications that a vehicle underwent from the time of impact until the inspection by the experts.

Photographs 2.1: Differences in the Frontal Part between the Time of the Accident and the Eye Inspection

Photographs 2.2: Differences in the Position of the Backrest between the Time of the the Time of the Accident and the Vehicle Inspection

Photographs 2.3: Differences in the Left Doors between the Time of the Time of the Accident and the Inspection of the Vehicle

Picture 2.4: Maximum Speed Sign
(omitted from External Information)

Source General

- Specialised Bibliography
- General Bibliography
- All other material of interest

At this point, all relevant bibliographic information will be considered in order to carry out the accident reconstruction. Specialised bibliography includes Traffic Accident Reconstruction Manuals, regulations and scientific articles applicable to the case under study.

In addition, depending on the case analysed, it may be necessary to resort to specific or generic bibliography that applies to particular details, such as fractomechanics manuals, studies on the impact of alcohol on driving, among many other sources that the expert must consult in order to adequately fulfil his task.

2.3 Information Analysis

The accident reconstruction expert will meticulously analyse and organise all the information gathered on the case. This includes photographs and/or videos of the collision scene, police reports, statements from the parties involved and witnesses, expert reports, the condition of the vehicles after the impact, characteristics of the accident site, medical reports, and any other data relevant to the analysis.

During this process, the expert will structure the information, looking for patterns, identifying possible inconsistencies or contradictions, and evaluating each element with a critical approach.

This detailed analysis will not only give you a better understanding of the facts, but will also provide you with the basis for formulating a solid hypothesis describing how the accident may have unfolded.

2.4 Hypothesis

Once the above analysis has been completed, the specialist will focus on integrating all the pieces of information obtained before, during and after the collision to understand the dynamics of the accident in a comprehensive manner. This process will allow him/her to identify the possible causal factors that contributed to the event. From this assessment, the expert will proceed to formulate various hypotheses which, in his professional opinion, may explain the different versions of the accident, since it is common for the parties involved to present contradictory accounts. These hypotheses usually consist of a detailed description of the sequence of events, from the moments leading up to the development of the accident, and should be consistent, taking into account all available evidence, such as vehicle damage, road conditions, estimated speeds and driver behaviour.

Each hypothesis will be further processed, allowing the expert to assess its degree of validity and to explain to stakeholders the robustness of each hypothesis. This interpretation will be supported by objective and indisputable scientific principles, and at the end of the proposed process, a clear and substantiated view of the accident will be obtained.

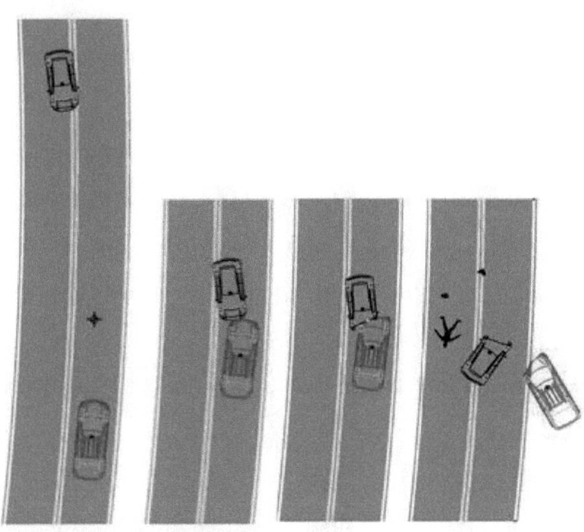

Figure 2.3: Example of the Sequence of Events of a Hypothesis

2.5 Mathematical Model

On the basis of the hypotheses formulated and the objectives defined by the interested party, the expert will proceed to develop a mathematical model to represent them, which will be composed of a set of equations describing the dynamics of the accident according to the hypotheses formulated, based in particular on the principles of conservation of energy and quantity of movement, and complemented by equations of the dynamics and kinematics of bodies, among others.

However, as is well known by specialists in road accident reconstruction, the values of all the variables involved are rarely known with precision. Moreover, these mathematical models often have more variables than equations, which generally leads to multiple or indeterminate solutions, adding complexity to the process and requiring a rigorous treatment that the proposed methodology allows to solve.

In the conventional solution, the expert addresses this drawback by assigning estimated values to some of the variables, with the aim of obtaining a given system of equations (where the number of equations is equal to the number of unknowns). However, this strategy is left to the discretion of each expert, which can lead to a solution with a considerable risk of not conforming to reality and/or being inconsistent.

To remedy this drawback with scientific rigour, the methodology that has motivated the writing of this work has been developed, which, unlike the conventional approach, does not assign estimated values to some of the unknowns, but adds a (functional) equation to maximise or minimise as appropriate to the case to be solved.

Details on the approach and implementation of this methodology are described in Chapters 3 and 4.

2.6 Criteria Archive

Through the proposed methodology, the mathematical model is solved, identifying all the feasible solutions, which are stored in a file. Subsequently, these solutions are ordered from best to worst according to the criteria defined by the expert, and mathematically captured in the functional.

This file, called the "Criteria File", will serve as the basis for identifying the most appropriate solution, ensuring that it is consistent with the available information.

2.7 Analysis of the Results

The information contained in the Criteria File will be analysed by the expert, considering all those circumstances that the mathematical model does not reflect, either because of its inherent limitations or because they were not contemplated in the initial analysis.

Once the Criteria File has been analysed, the expert must make a decision, based on the following criteria: when the results are completely consistent with the available information and the expert's judgement, the hypothesis will be accepted as the most likely explanation of the facts; otherwise, it should be discarded or at least reformulated, restarting a new cycle from the most appropriate point (for example, requiring additional or more precise data; analysing the information from another approach, redefining the parameters of interest, the hypotheses, the objectives and/or the mathematical model). This decision is called a Satisfactory Decision.

To illustrate a possible Satisfactory Decision, suppose the expert observed that the coefficients of friction between the vehicle and the pavement were unusually low. In this case, he could infer that the pavement was wet or contaminated with oil at the time of the accident. To confirm this hypothesis, he would consult weather records, interview witnesses and/or make on-site measurements of the coefficients of friction, rather than relying solely on literature values.

2.8 Validated Hypothesis

The process described in the previous point will be repeated iteratively until a solution is found in which the parameters used - after a suitable convergence process - are fully consistent with the available evidence and the expert's criteria. In this way, the hypothesis will be confirmed and the values of the parameters sought will be obtained. It can then be stated that the hypothesis has been validated and that a solution has been obtained that is an excellent and reliable representation of the events that occurred during the accident. This solution must be thoroughly documented and will be included in the corresponding expert reports.

CHAPTER 3

OF THE MATHEMATICAL MODEL TO THE ANALYSIS OF RESULTS

3.1 Introduction

This chapter describes the computational implementation of the Mathematical Model to obtain the feasible solution space, which will be structured in the Criteria File. Subsequently, we will proceed to the analysis of the results contained in this file. Figure 3.1 highlights the three blocks discussed in this chapter.

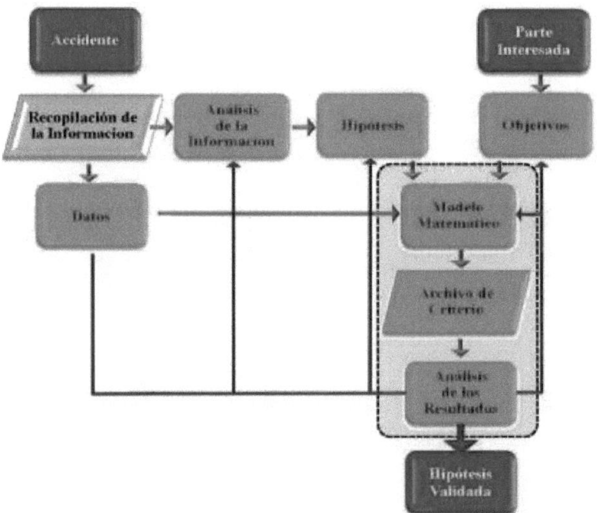

Figure 3.1: Reconstruction Process (blocks covered in this chapter)

3.2 Mathematical Model

It will consist of a set of equations representing the accident dynamics, together with a functional equation to be maximised or minimised. The solution of this type of model is analogous to that of a multivariable optimisation problem with constraints.

3.2.1 Limitations of the Analytical Solution

Although, from a theoretical point of view, the analytical solution of the mathematical model is a valid alternative, in practice, and especially in the reconstruction of road accidents, it has a number of significant limitations:

- Optimisation in these cases usually involves solving systems of non-linear equations involving trigonometric functions.
- The analytical resolution of these systems requires dealing with systems of non-linear differential equations, which is only feasible by numerical methods, which often present convergence problems.
- The equations (functions) must be continuous and derivable, which is not always the case for the variables involved in accident reconstruction, where occasionally some of the variables occur in discrete jumps.
- The solutions obtained analytically also do not clearly express the sensitivity of the result to variations in its environment.

For example, when solving analytically a univariate problem for the functionals A and B, we obtain that the minimum value of each is 2, which is achieved when the independent variable takes the value of 0, i.e., although both functionals have the same values, we observe that the environments around the minimum are significantly different. Since the anahetic solutions do not provide information about these environments, we must resort to the discrete solutions.

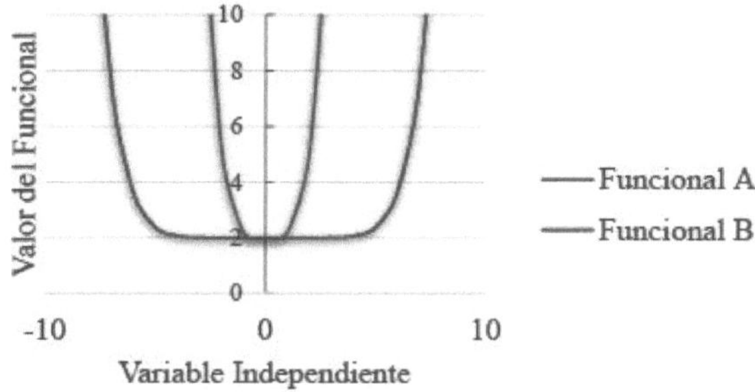

Figure 3.2: Environments around the Optimum Point

3.2.2 Discrete solution

Due to the limitations of analytical optimisation, a better alternative is to solve the mathematical model by means of a particular discrete optimisation process.

This approach not only allows the identification of the optimal point, but also provides information about its environment.
For example, Figure 3.3 shows the discretised results of the functionals shown in Figure 3.2.

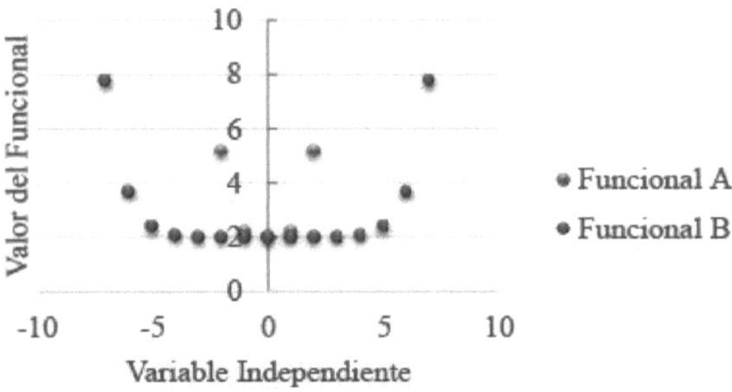

Figure 3.3: Environments around the Optimal Point

In this figure it can be clearly observed that the minimum value of functional B remains practically stable between -4 and +4, while, in functional A, the optimum varies considerably outside the range of -1 to +1. This ability to analyse the environment provides greater versatility to define the feasible solution that best fits the information available in each case to be solved.
The main advantage of the discrete treatment, as proposed in this text, lies in its ability to generate not just one, but a set of feasible solutions. By being compiled in an orderly fashion in the Criterion File, these solutions allow the expert to analyse them in a very simple way, and to make a satisfactory decision, based on a combination of the mathematical results and his professional judgement. Through this iterative process, the expert will be able to determine with a high degree of certainty how the facts unfolded.

3. 3Implementation

3.3.1 Definitions

- The following definitions shall be used in the mathematical model:

X: Design or Independent Variables (are those that will be varied within a certain pre-specified Range)

Y: Dependent Variables (are determined from the design variables).

Z: Functional (represents the equation to be maximised or minimised, as defined by the expert).

- The mathematical model will be constituted by:

I System of Equations: **Y** = f (**X**)

| Functional: **Z** = f (**X**, **Y**) Max o Min

- The mathematical model will be solved on the basis of the methodology proposed in the following points.

3.3.2 Preparing for Resolution

It is important to note that there is no single way to approach the mathematical modelling of a particular accident, as the design variables to be considered will depend on the judgement of the expert. In an extreme case, all the parameters of interest could be treated as design variables, which would imply including the corresponding conditions to ensure the feasibility of the solution. However, it is essential to stress that the solution that best explains the accident will be the same, regardless of how the mathematical model is structured in terms of design variables and dependent variables.

Once the Mathematical Model has been defined, proceed as follows:

1) Identify hard data, i.e. data about which there is absolute certainty as to its value.

2) Identify the design variables, which are those whose value we do not know exactly (x#).

3) For each of the design variables, we will define a reasonable range of values within which the real value could be found, for this we will define its minimum value ($x\#_{min}$) and its maximum value ($x\#_{max}$). Given that, within the iterative process, these design variables will be modified in a discrete way, it is necessary to specify their corresponding variation step ($x\#_{paso}$).

4) Arrange the equations in such a way that the dependent variables are a function of the design variables. In case any dependent variable is also a function of another dependent variable, the latter must be calculated first.

5) Generate one or more conditions to identify whether the combination being

processed in this cycle is feasible or not.

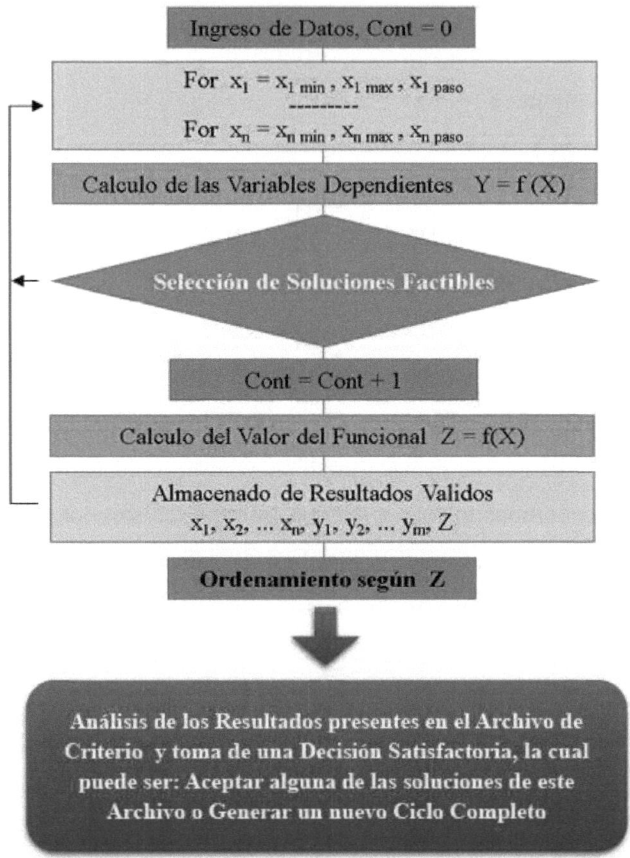

Analysis of the Results in the Criteria File and making a Satisfactory Decision. Criteria File and making a Satisfactory Decision, which can be may be: Accepting one of the solutions in this File or Generate a new Complete Cycle

Figure 3.4: Resolution process

3.3.3 Programming

With the mathematical model and the information prepared in the previous point, we will proceed to develop a computer programme following the guidelines established in Figure 3.4, in which the following steps can be visualised:

1) The hard data is entered and the feasible solution counter (Cont) is reset.

2) An iteration structure (loop) is programmed where each of the design variables is computed in each cycle from the minimum Hmite to the maximum Hmite according to the preset step.

3) With the hard data and the design variables, the dependent variables will be computed.

4) The process continues to check if the solution obtained in this iteration is feasible. If it is not, the iterative process is restarted by advancing one step in the corresponding design variable.

5) If the solution is feasible, the feasible solution counter is incremented by 1 (one) and the value of the functional is calculated.

6) The solution reached in this iteration is stored in a file, which includes at least the values of all variables and the corresponding functional. If this solution is not the last one, the iterative process is restarted by advancing one step in the corresponding variable.

7) Once all the all the

possible combinations, all the stored solutions are sorted from best to worst, according to the criteria established in the functional. We call this file the Criteria File.

3.4 Analysis of the Results

The expert will proceed to analyse the results present in the criteria file, starting with the first solution in the list, which is assumed to be the best. In this analysis special emphasis will be put on the consistency of each feasible solution with the available information. When this compatibility is complete, the solution will be accepted as the most probable explanation of the facts and consequently the Hypothesis will be Validated. Otherwise, it should be discarded or at least reformulated, restarting a new cycle from the point that the expert considers more appropriate (for example, requiring additional or more precise data; analysing such information from another approach or redefining: the parameters of interest, the hypotheses, the objectives and/or the mathematical model).

This iterative process, which seeks to construct a hypothesis that coherently and scientifically explains all the events of the accident, should be initiated at the earliest stages of the investigation. Through successive iterations and constant evaluation of the evidence, the hypothesis will be refined until a solid and consistent explanation is achieved. The process concludes when a hypothesis is identified that is consistent with all available evidence and the unbiased judgement of the expert.

CHAPTER 4

CASE STUDY

4. 1Case Presentation

To clarify the use of the methodology presented in Figure 2.1, a simple theoretical case study will be considered, in which a truck impacts a stationary pick-up truck that is in its path.

4.2 Objective

It is assumed that the Interested Party has defined that the primary objective in this case is to *determine the speed of the truck in order to know whether it exceeds the permitted limits*.

4.3 Compilation of information

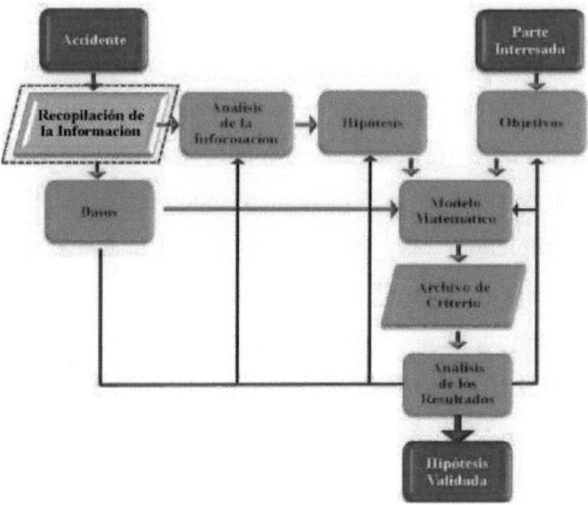

Figure 4.1: Collection of Information

After the accident, the expert shall collect information from external, internal and general sources as detailed in section 2.2.2.

In this case study it is assumed that the information collected has been as follows:

- from External Sources:

- Sketch of the police expertise, which shows the tracks left by the vehicles and their respective lengths.
- The court file contains photographs of both vehicles, showing their respective damage.
- Witness and driver statements are in the court file.

- from Domestic Sources:

- The site was inspected by the expert (weeks after the accident).
- The vehicles were inspected by the expert (weeks after the accident).

of General Sources:

- Reconstruction Manuals were used to identify the formulations associated with the accident events and to estimate some of the related data. In addition, vehicle owner's manuals were consulted to obtain information on vehicle weights, wheelbases and wheelbases.

Since the objective of the presentation of this case study is to explain in a simplified way the proposed methodology and considering that the Stakeholder's stated objective is only to determine whether the speed of the truck was excessive, it is not necessary to complete the Haddon matrix.

In real cases, completing this matrix helps to organise the information and ensures that no relevant data is omitted, which can be crucial in identifying specific details of the accident.

4.4 Analysis of Information

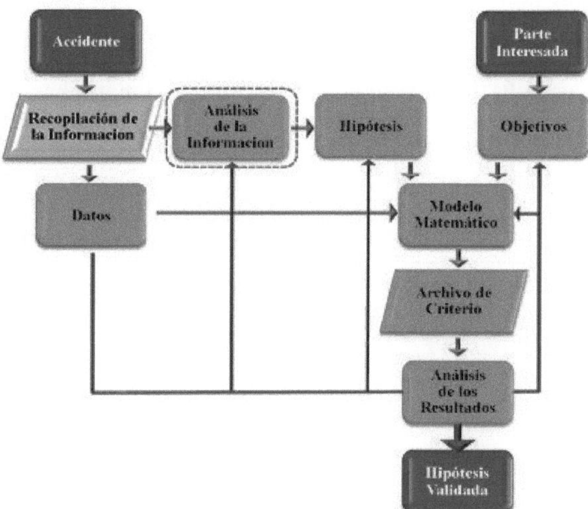

Figure 4.2: Information analysis

On the basis of the information gathered in the previous point, we proceed to its analysis. In this case study, the following conclusions will be considered to have been obtained (to be verified at the appropriate time in this process):

- The distance between the longest tracks on the police sketch corresponds to the track of a truck and that the truck was moving forward on a braking trajectory.
- The distance between the other tracks on the police sketch corresponds to the wheelbase of a pickup truck, indicating that it moved laterally in the same direction as the truck.
- The straight shape of the tracks and their parallelism are evidence that the van was towed by the truck while its speed was zero or extremely low.
- Witness and driver statements are consistent in mentioning that the van was stopped when it was hit by the truck and that the truck braked before hitting it.
- The above statements are consistent with the shape and direction of the fingerprints present in the court file.
- The photographs in the file reflect the same damage observed during the visual inspection of the vehicles.
- The damage observed on the vehicles confirms that the front of the truck hit

the left side of the pick-up with its front end.

- The inspection of the site revealed that the road is straight and the carriageway is asphalt, that it is in good condition and that it has no slope (useful data for the equations of the mathematical model).
- The reconstruction manuals allowed the identification of the formula to be used and the typical friction coefficients between wheels and pavement.
- The user manuals made it possible to establish the masses (weights) of the vehicles involved in this accident.

4.5 Hypotheses

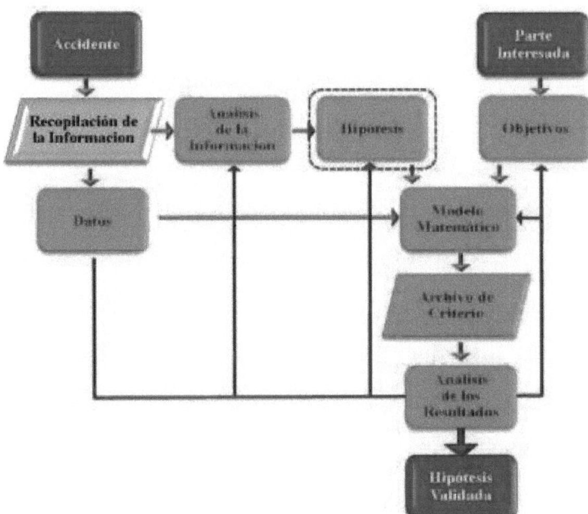

Figure 4.3: Hypotheses

The hypothesis put forward in this case was as follows: *the accident developed according to the sequence of events described below.*

a) Event 0 - Start of Accident

The truck driver notices that the van is in his way, so he starts braking while driving at speed v_{C0}. Since the objective is to determine this speed (Point 4.2), v_{C0} is the most important parameter in this reconstruction.

Based on the statements and the rest of the evidence, it is considered that the speed of the pick-up is 0 (zero), establishing T_0 as the starting time of the accident dynamics.

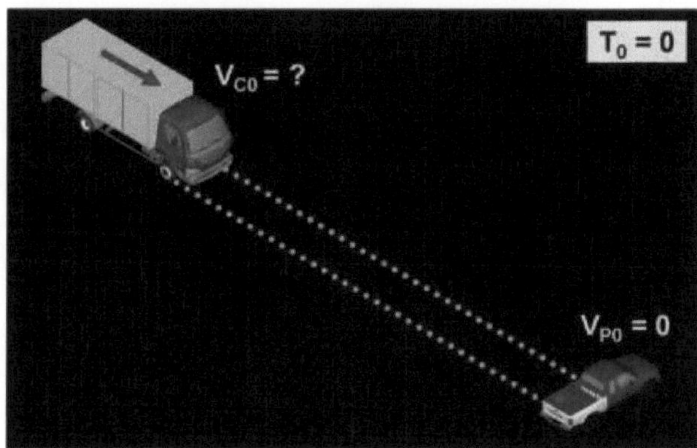

Figure 4.4: T0 - Start of Accident

b) Event 0 to 1 - Truck Braking

This event considers the trajectory of the truck during braking prior to impact. This trajectory is evidenced by the footprint found in the police report.

The speed of the truck during this trajectory will be reduced until the moment of impact.

The van remains stationary throughout this event which takes place between times T0 and T1.

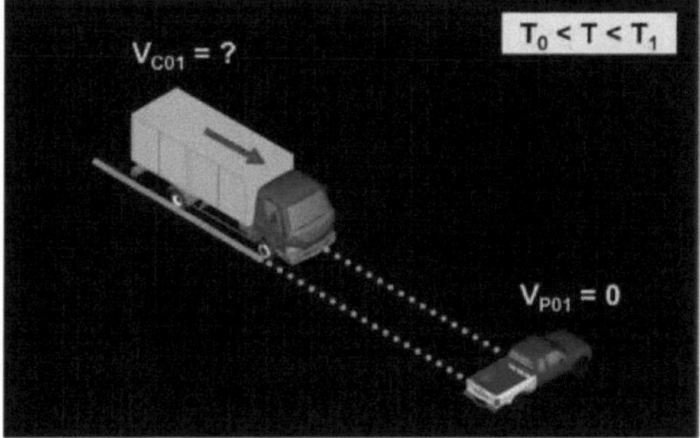

Figure 4.5: T0 < T < T1 - Truck Braking

c) Event 1 - Impact

The truck hits the pick-up at speed V_{ci}. At this point, the pick-up remains stationary, as it has been throughout Event 0 to 1.

Immediately after impact, both vehicles couple and start to move together at speed V_1.

It is established that the impact occurs at the time

called T1.

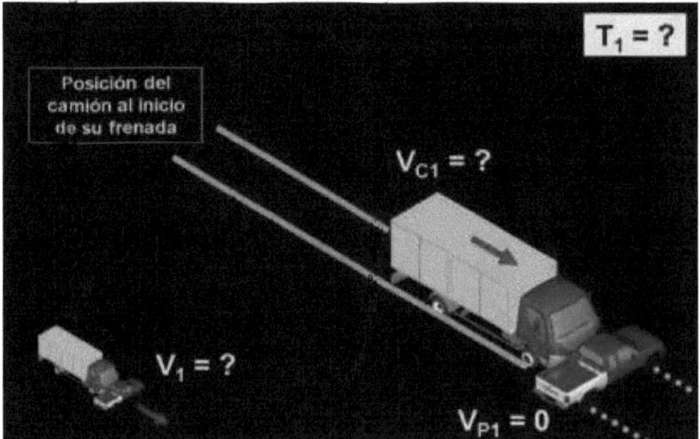

Figure 4.6: T1 - Moment of Impact

d) Event 1 to 2 - The Truck Drags the Van

During this event, the truck is braking and leaving the track recorded by the police. In addition to this footprint, the pick-up truck is also leaving a footprint due to its lateral dragging.

The speed of both vehicles decreases gradually from the moment of impact until they come to a complete stop.

This event takes place between times T1 and T2.

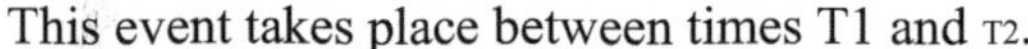

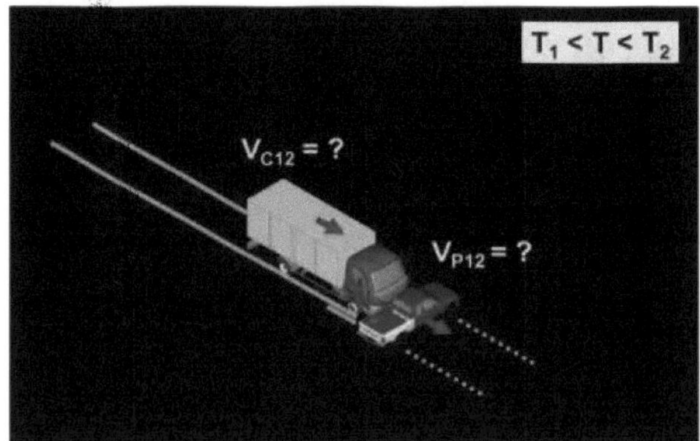

Figure 4.7: T1 < T < T2 - Truck towing the van

e) Event 2 - End of Accident

Both vehicles reach the final rest position at the time called T2.

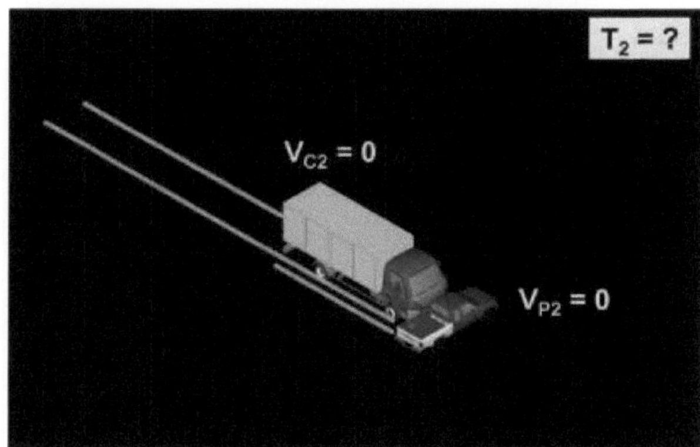

Figure 4.8: T2 - Final Positions in the Accident

This last event ends the sequence of events that define the hypothesis that will be modelled and eventually validated in the following points.

4.6 Mathematical Model

4.6.1 Introduction

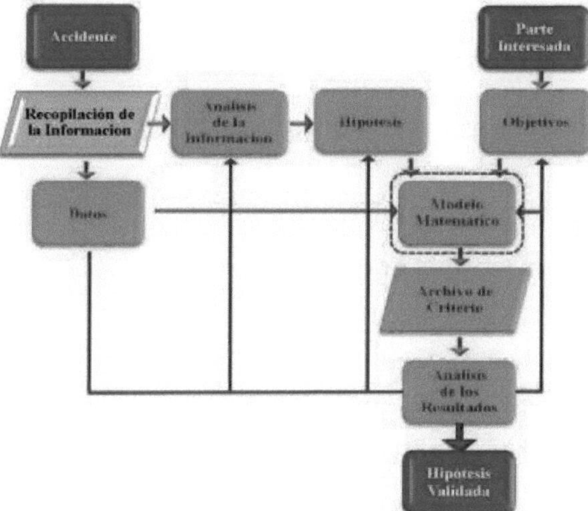

Figure 4.9: Mathematical Model

At this stage of the process, we are now in a position to elaborate the mathematical model, which should adequately reflect the hypothesis and meet the established objective.

The following tables present a summary of the sequence of events and their main parameters:

#	EVENT
0	Onset of the Accident
0 a 1	Truck Braking
-1	Pre-Impact Moment
1	Impact
-1	Post-Impact Moment
1 a 2	The Truck Drags the Van
2	End of Accident

Table 4.1: Sequence of Events

#	TIME	SPEED	
		Truck	Pick-up
0	To	Vco	VP0 = 0
0 a 1	$T_0 < T < T1$	VC01	VP01 = o
-1	~ T1	VC1	VP1 = 0
1	T1	VC1 > V > V1	VP1 < V < V1
-1	~ T1	V1	V1
1 a 2	$T1 < T < T_2$	VC12 = VP12	VP12 = VC12
2	T2	VC2 = 0	VP2 = 0

Table 4.2: Sequence of Events and their Main Parameters

The following parameters will be used to formulate the mathematical model:

V Truck Speeds:

V co:at the start of braking (TO)

V C01: during the pre-impact trajectory

V C1:upon impact of the van

V 1: at the start of the truck's towing (T1)

V C12: during the trailing trajectory

V C2:at the end of the accident (T2)

V Pick-up truck speeds:

V P0:at start of truck braking (TO)

V P01: during the pre-impact trajectory of the truck

V $_{P1}$: on impact

V $_{1}$: at the start of the trailing trajectory ($_{T1}$)

V $_{P12}$: during the trailing trajectory

V $_{P2}$: at the end of the accident ($_{T2}$)

V Vehicle masses: m_c: mass of truck

m_p: mass of the van

V Friction Coefficients:

f_c: truck-road

f_P: truck-roadway

V Distances travelled from the:

d_{C01}: Start of braking to Impact (truck)

d_{C12}: Impact to Detention (truck) d_{P12}: Impact to Detention (van) d_{12}: Impact Until Stopped (both vehicles)

V Deformation Energies:

E ($_{DD}$): Energiadedeformationofboth vehicles calculated on the basis of damages.

$E_D(V)$: Energiadedeformationofboth vehicles calculated on the basis of the speeds.

$E_\#$: Kinetic or frictional energies. The sub-indices # represent the vehicle and the corresponding event (see section 4.7).

- Others:

g: Gravity Acceleration

4.6.2 Definition of the Mathematical Model

The equations governing this accident are essentially based on the principles of energy conservation and quantity of motion. The equations corresponding to each event are presented below.

a) Event 0 - Start of Accident (T = 0)

This event has no equation to represent it. It is only known that the truck was moving at a speed called v_{C0}, whose value is unknown, and that the pick-up was stopped, that is to say, that the truck was moving at a speed called v_{C0}, whose value is unknown:

$$v_{C0} = ? \quad v_{P0} = 0$$

b) Event 0 to 1 - Truck Braking ($T0 < T < T1$)

From the beginning to the end of this event the pick-up was stopped, therefore:

$$v_{P0} = v_{P01} = v_{P1} = 0$$

The truck at the beginning of this event had a kinetic energy that can be calculated as:

$$E_{C0} = 0,5 \,.\, m_C \,.\, v_{C0}^2$$

During its braking, the truck lost part of its kinetic energy, which can be computed as:

$$E_{C01} = m_C \,.\, g \,.\, f_C \,.\, d_{C01}$$

The truck at the end of this Event (i.e. immediately before the impact) had the following kinetic energy:

$$E_{C1(-)} = 0,5 \,.\, m_C \,.\, v_{C1}^2$$

By the principle of energy conservation it can be established that the energy balance must be zero, therefore:

$$E_{cho} - E_{choi} - E_{ci(-)} = 0$$

Although the required computer program can be carried out using all the above equations directly, in order to simplify and reduce the number of lines of code, it is advisable to consolidate these equations into a single equation to determine the required parameter in the target:

$$V_{co} = (V_{ci2} + 2\, g\, f_C\, d_{coi})^{0,5}$$

c) Event 1 - Impact

To model the impact between the truck and the pick-up, the principle of conservation of the quantity of motion can be used, which says that the quantity of motion of the vehicles before the impact must be equal to the quantity of motion of the vehicles after the impact.

$$m_C \,.\, v_{C1} + m_P \,.\, v_{P1} = m_C \,.\, V1 + m_P \,.\, v_1$$

Rearranging the previous expression and as V_{PI} (velocity of the truck before impact) is zero, we obtain:

$$V_{ci} = (m_e + m_p)\, V_i / m_e$$

Furthermore, in this event, equations derived from the ene^a conservation principle can also be posed, since the pre-impact enemas must be preserved after the impact.

The kinetic energKi of the system at the instant before impact is given by the following equation:

$$E_{ei(-)} = 0.5\ .\ m_e\ .\ V_{ei}^2 \text{ (the van does not contribute)}$$

As after the impact both vehicles are coupled together their velocities will be equal, therefore the kinetic energy of the system immediately after the impact will be:

$$E_{epi(+)} = 0.5\ .\ m_e\ .\ V_i^2 + 0.5\ .\ m_p\ .\ V_i^2$$

Furthermore, during the impact, deformation of the vehicles also occurs. In this case study we assume that - from the analysis of the possible photographs and the information collected during the inspection of the vehicles - the truck does not show significant deformations, but the pick-up does. On the basis of these deformations and using the corresponding stiffness coefficients it can be calculated that the energy lost in deformations during the impact has been:

$$E_{D(D)E} = 255 \text{ kNm}$$

In order to offer a practical approach to accident reconstruction, this chapter focuses on the fundamental aspects of the efficient methodology, leaving aside more detailed calculations such as the energy dissipated in deformations.

As the ene^a must be conserved during this event, the corresponding equation will be:

$$E_{ci(-)} - E_{CPI(+)} - E_{D(D)} = 0$$

In this case it is also possible to implement the code using all the equations presented, but for simplicity and efficiency, they have been condensed into a single equation:

$$E_{D(V)} = 0{,}5\ m_e\ V_{ci2} - 0{,}5\ (m_e + m_p)\ V^*$$

The values of ED(V) and ED(D) should be the same as they represent the same physical phenomenon, but they have been calculated by two different methodologies.

d) Event 1 to 2 - The Truck Drags the Van

During this event both vehicles move as a single mass, so their velocities will be equal:

$$v_{P12} = v_{C12} = v_{12}$$

During this event, both sails moved the same distance:

$$d_{P12} = d_{C12} = d_{12}$$

As previously stated the kinetic energy of the system at the beginning of this event (immediately after the impact) is:

$$E_{CP1(+)} = 0.5\ m_C \cdot v_1^2 + 0.5\ m_P \cdot v_1^2$$

Due to the braking during this event the truck lost the following amount of energy:

$$E_{C12} = m_C \cdot g \cdot f_C \cdot d_{12}$$

Due to its drag the pick-up lost the following amount of energy:

$$E_{P12} = m_P \cdot g \cdot f_P \cdot d_{12}$$

At the end of the event, both vehicles came to a standstill and consequently their kinetic energy was zero. With this consideration and taking into account the principle of conservation of energy it can be established that:

$$E_{CP1(+)} - E_{ci2} - E_{Pi2} = 0$$

Although in this case it could also be programmed using all the above equations, in order to simplify the code, we have worked with these equations to obtain a single equation:

$$V1 = [2\ g\ d_{i2}\ (m_e\, f_C + m_p\, f_P\ ) / (m_e + m_p)]\ 5^{0,}$$

e) Event # 2 - End of Accident

This event has no equation to represent it. It is only known that both vehicles reach their final resting position, therefore:

$$v_{C2} = 0$$

$$v_{P2} = 0$$

4.7 Preparation for the Resolution

1) Identifying hard data

They are assumed to be known with a minimum uncertainty a:

g: **9.81** m/s^2

mP: **1930** kg (mass in running order obtained from the owner's manual plus the mass of the driver)

2) Identify the design variables, which are those whose value we do not know exactly (x#).

Among the design variables that have been identified are the friction coefficients between the vehicles and the road (fCand fp), since, although a statistical value is available, the exact values for this case are unknown. Also included are the distances travelled by vehicles (dcoi and di2.) as evidenced by the footprints on the pavement. Although these are defined in the police expertise, it is assumed that they may have some measurement error.

Finally, the mass of the truck is included. Although its model is known, it is assumed that in the information surveyed the mass of the load transported was not defined, therefore, the expert is not in a position to propose a reasonable estimated value.

3) Range of values for the Design Variables.

For each of these variables a reasonable minimum and maximum value is set at the discretion of the expert, and the corresponding step is incorporated, which can be as refined as it is possible to handle on our computer. The recommendation is that all combinations generated during the process should not exceed a value of 30,000 so that the runs are dynamic and the criteria files are not so large. If the feasible solution found is too sensitive to any of the design variables, a new iterative process can be generated after analysing the criteria file, reducing the step as well as the range.

Design Variable	Minimum Value	Maximum Value	Passage
fc	0,40	0,80	0,02

fp	0,40	0,80	0,02
dcoi	29 m	31 m	0,5 m
di2	13 m	15 m	0,5 m
mc	3200 kg	8200 kg	500 kg

Table 4.3: Design Variables to be Considered

4) Ordering of equations:

The equations presented in the previous points are then transcribed, but arranged in such a sequence that all the variables are a function of known or already determined parameters:

$$V1 = [2\ g\ di2\ (me\ f_e + mp\ f_p\ )\ /\ (me + mp)]\ 0{,}5$$

$$Vci = (me + m_P\ )\ Vi\ /\ m_c$$

$$V_{co} = (Vci2 + 2\ g\ fe\ deoi)$$

$$ED_{(V)} = 0{,}5\ m_e\ Vci2 - 0{,}5\ (mC + m\ )_p \qquad 2$$

5) Generate feasibility conditions:

In this case, it can be established that the deformation energy calculated from the vehicle deformations (ED(D)) should coincide with the value obtained according to the energy conservation principle applied at the moment of impact (ED(V)). Since the energy estimated from the deformations can be imprecise, a solution will be considered valid if it is within a range of 5% above or below the expected value.

$$\mathbf{0{,}95\ .\ ED(D)E < ED\ (V) < 1{,}05\ .\ ED(D)E}$$

As the E ( )$_{DDE}$ was calculated at 255 kNm the previous equation can be written as:

242 kNm <^{E} $_{D(V)}$< 268 kNm

6) Functional:

The functional incorporated in the mathematical model tries to minimise the errors between the expected values, defined according to the expert's criteria, and the values obtained through the resolution of the mathematical model.

Variable	Value according to Criterion
fcE	0,6
fPE	0,6
dcoiE	30 m
d12E	14 m
ED(D)E	255 kNm
mcE	3200 to 8200 kg

Table 4.4: Expert Expected Values

Therefore, the error functional to be minimised, in the ***i-th*** iteration, will have the following form:

$$Z = [(f_{ci} - f_{cE}) / f_{cE}]\ 2 + [(f_{pi} - f_{pE}) / f_{pE}]^2 +$$
$$+ [(dco_i - dc01_E) / dc0_1E]^2 + [(d12i - d12E) / d12E] + [(d12i - d_{12}E) / d12E] +$$
$$[(d12i - d12E) / d12E]^{\ 2}$$
$$+ ((E_{D(V)\,i} - E_{D(D)E}) / E_{D(D)E})^{\ JI\ 2}$$

4.8 Programming

The mathematical model is programmed on the basis of the seven items of point 3.3.3. and the code is obtained, which can be seen in Annex A: Calculation Program.

From the execution of this software it was possible to obtain the corresponding Criteria File, whose best feasible solutions are presented below:

ErrorVco		Vci	Vi	ED(V)fc	fp	dcoi	di2	mc
0.	*000100.*	*374.*	*146.*	*22550.*	*600.*	*6030.*	*014.*	*03200*
0.	00099.	974.	146.	22550.	600.	6029.	514.	03200
0.	000100.	774.	146.	22550.	600.	6030.	514.	03200

Table 4.5: Criteria File - Best Feasible Solutions

NOTE: If no solutions are found when executing the program, first check the lines of the code. If these are correct, the program should be examined in its entirety. If the problem is not found there, it is necessary to check the data, both the hard values and the design variables. If still no solutions are obtained, the hypothesis should be discarded or reformulated, and the process restarted from the point that the expert identifies as the most appropriate.

4.9 Analysis of the Results

The analysis of the criteria file consists of analysing the consistency of the results obtained based on the Expert's expectations (criteria). The recommended steps to carry out this analysis are described below.

- Range Verification

The first check to perform is to verify that the Design Variables, determined by the program, are within the preset range.

Design Variable	Minimum Value	Maximum Value	Result
fc	0,40	0,80	**0,60**
fp	0,40	0,80	**0,60**
dcoi	29 m	31 m	**30 m**
di2	13 m	15 m	**14 m**

Table 4.6: Results of the Design Variables Vs. their Rank

In this case it is observed that the design variables are all within the preset range. If any of them are at an extreme, the range should be extended and a new run should be performed to obtain an updated criteria file. This process should be repeated until the above condition is met.

Among the results obtained are also the values of the mass of the truck:

Design Variable	Minimum Value	Maximum Value	Result
mc	3200 kg	8200 kg	**3200 kg**

Table 4.7: Results of mC Vs. its Rank

This variable is a special case because, although it is at the extreme end of its range, the extremes are reliable data: they represent the truck both in its unloaded and fully loaded state. Therefore, it is not necessary to extend the range, since the value obtained is the minimum physically possible.

- Consistency Check

In this step, the extent to which the results are in line with the expert's judgement should be verified:

Design Variable	Expected Value according to Criteria	Result
fc	0,6	**0,60**
fp	0,6	**0,60**
dcoi	30 m	**30 m**
di2	14 m	**14 m**
ED(D)	255 kNm (ED(D)E)	**255 kNm** (ED(V))

Table 4.8: Expected Values of Variables Vs. Results

In the case study analysed, the coincidence has been total, a situation favoured in some way by the fact that the errors between the value defined according to the expert's criteria and the result obtained by the software, presented in the criteria file, have been minimised.

If these values do not coincide, it is left to the discretion of the expert to make a so-called 'Satisfactory Decision', which will depend on which variables are in discrepancy. In order to clarify this treatment, it could be considered that the friction coefficient determined for both vehicles was too low in relation to the expert's expectations. These values could indicate that the road was wet or contaminated. The expert should therefore try to confirm this analytical result with reality, for example by consulting the meteorological service or drivers and witnesses, and then act on this new information.

Additional Identifications

The optimal solution, which has the smallest margin of error, confirms the estimated values for the friction coefficients, the distances travelled by the vehicles and the deformation energy. However, when the problem was posed, the expert did not know the mass of the truck load and, therefore, the total mass of the truck, and consequently did not have an estimated value for them. For this reason, the behaviour of the system was analysed considering a range of variation of the truck mass, from the no load condition to the full load condition. The best result indicates that the total mass of the truck at the time of the accident must have been 3200 kg, which clearly indicates that the truck must have been unloaded. Consequently, from the analysis of the criterion file, it is concluded that, for the truck to have shown the behaviour observed in the accident, it must have been running unladen.

With this information, the expert should proceed to verify the accuracy of this conclusion, using methods such as consulting the driver, interviewing witnesses or reviewing the truck's documentation, in order to validate the proposed hypothesis. If these enquiries reveal that the truck was loaded to half capacity (e.g. the total weight to be considered should have been 5300 kg), this new information is incorporated and the process is restarted to find the new set of feasible solutions, shown in Table 4.9.

ErrorVco		**Vci**	**Vi**	**ED(V)fc**	**fp**	**dcoi**	**di2**	**mc**
0.	*01196.*	*766.*	*849.*	*02440.*	*640.*	*6030.0*	***15.0***	*5300*
0.	01296.	366.	849.	02440.	640.	6029.	515.	05300
0.	01297.	166.	849.	02440.	640.	6030.	515.	05300

Table 4.9: Criteria file - Mass of 5300 kg truck

From the analysis of this new matrix it can be identified that the value of the distance travelled post impact corresponds to the upper limit of its range, consequently it should be extended and a new iteration should be performed, repeating the process as many times as necessary until the hypothesis can be validated.

In short, the expert will thoroughly evaluate all the results obtained and, on the basis of his or her professional judgement, make a satisfactory decision. When the results are completely consistent with the available information and his expert judgement, the hypothesis will be accepted as the most probable explanation of the facts; otherwise, it should be discarded or at least reformulated, restarting a new cycle from the most appropriate point (for example, requiring additional or more precise information; analysing such information from another approach, redefining the parameters of interest, the hypotheses, the objectives and/or the mathematical model).

4.10 Validated Hypothesis

The process described in this example will be repeated iteratively until a solution is found in which the data used - after a suitable convergence process - are fully consistent with the available evidence and the expert's judgement. In this way, the hypothesis will be confirmed and the values of the parameters sought will be obtained.

In this case study, as the results obtained in Table 4.5 (after confirming that the truck was unloaded) are fully compatible with the available information and with the expert's judgement, the sequence of events defined in the hypothesis can be considered correct, and therefore we can say that the hypothesis has been validated and is the best representation of the events that occurred during the accident.

Also, in this example, the objective set by the Interested Party, which is to determine the speed of the truck in order to verify whether it exceeds the permitted limits, has been fulfilled.

CONCLUSION: The lorry was travelling at 100 km/h, so it is clear that it significantly exceeded the speed limit for lorries (80 km/h). This infringement represents a significant risk factor which contributed decisively to the dynamics of the accident and the seriousness of its consequences.

4.11 Other Treatments

Given the recurrent interest of the parties involved in knowing the maximum or minimum speed of the vehicles involved in an accident, a specific section (Annex C: Other Treatments) has been devoted to address this issue.

4.12 Concluding Remarks

The wide range of variables that influence an accident makes it unfeasible to foresee all possible combinations in this text. For this reason, the results file, called the 'Criterion File', is designed to be supplemented by expert judgement to ensure the validity of its conclusions.

In short, the reconstruction of an accident requires a holistic approach that integrates the expert's experience with the quantitative results provided by the mathematical model. This is why the synergy proposed in the methodology described in this text is indispensable to obtain a realistic and scientifically grounded representation of the event.

Conclusions

5.1 Innovative Methodology

This methodology offers a new perspective for the reconstruction of traffic accidents, combining expert knowledge with simplified mathematical models. Through an iterative process, accurate and reliable solutions are obtained, avoiding the subjectivity inherent in traditional methods.

5.2 Versatile Approach and Multiple Applications

The versatility of this methodology has led it to be applied in various fields, from space engineering to the food industry. Its capacity to solve complex problems and optimise systems or processes has been demonstrated on numerous occasions. Among others, it has been used for:

- Minimising the mass of satellite structures
- Minimise the rolling masses of a satellite.
- Optimising glass production
- Improving the efficiency of food production processes

5.3 Case Study

To illustrate the application of this methodology, a practical example has been presented, which, due to its simplicity, could have been solved in an analytical way, but its main objective has been to show in an executive way how the proposed methodology is implemented, highlighting its efficiency and its capacity to tackle much more complex problems without great additional efforts.

5.4 The Role of the Expert and the Criteria File

The expert plays a fundamental role in this process, defining the mathematical model, the ranges of the design variables and analysing the results and taking the so-called satisfactory decisions, which come from the combination between his expertise and the results obtained by means of the mathematical model, which are presented in the so-called Criterion File. This methodology, which combines the mathematical solution with the expert's judgement, is ideal because through an iterative process it answers the questions that become evident when analysing the criterion file, which is why it is excellent for detecting inconsistencies and exploring multiple scenarios.

5.5 Main benefits

- Accuracy: Mathematical models guarantee accurate and objective results.
- Versatility: The methodology can be adapted to a wide range of simple or complex

wide range of simple or complex problems.

- Objectivity: Reduces dependence on the subjectivity of the expert.
- Efficiency: Represented by the sum of the previous items and the fact that it allows solutions to be obtained quickly, even in complex cases.

5.6 Summary

This innovative methodology represents a significant advance in traffic accident reconstruction. Its versatile approach and its ability to solve complex problems make it a valuable tool for a variety of functions, and in particular traffic accident reconstruction.

ANNEXES

CALCULATION PROGRAMME

A. 1Introduction

Given its intuitive character and its free access, QBasic/QB64 has been the tool of choice for the implementation of the proposed algorithm. However, due to its simplicity, it can be easily replicated in other programming environments or even in spreadsheets such as Excel.

Furthermore, the simplicity of the algorithm and the structure of the code allow it to be easily adapted to model and solve any other type of traffic accident.

A.2 Code

Below is the code implemented in the first iteration of the case study (see Section 4.9):

```
'CASO de ESTUDIO.bas
'Todas las unidades no especificadas corresponden al Sistema Internacional

Dim Mat(38000, 10) 'Definición del tamaño de la matriz de soluciones

'INGRESO DE DATOS ******************************************

' Datos Duros ----------------------------------------------
g=      9.81    'aceleración de la gravedad
mp =    1930    'masa de la pick up - masa en orden de marcha

'Valores Esperados -----------------------------------------
EDDE = 255000 'energía consumida en la deformación de los vehículos
fcE =   0.6 'valor esperado del factor de fricción camión - pavimento
fpE =   0.6 'valor esperado del factor de fricción pick up - pavimento
dc01E = 30 'valor esperado de la distancia de frenado del camión pre impacto
d12E =  14 'valor esperado de la distancia de frenado del conjunto camión -
           pick up luego del impacto

' Puesta a Cero del Contador de Soluciones Factibles -------
Cont = 0

'FIN DE INGRESO DE DATOS ***********************************
```

```
' INICIO DEL PROCESO ITERATIVO ***************************
For fc = fcE - .2 To fcE + .2 Step .02
  For fp = fpE - .2 To fpE + .2 Step .02
    For dc01 = dc01E - 1 To dc01E + 1 Step .5
      For d12 = d12E - 1 To d12E + 1 Step .5
        For mc = 3200 To 8200 Step 500

          ' Calculo de las Variables Dependientes -------------------
          V1 = (2 * g * d12 * (mc * fc + mp * fp) / (mc + mp)) ^ 0.5
          Vc1 = (mc + mp) * V1 / mc
          Vc0 = (Vc1 ^ 2 + 2 * g * fc * dc01) ^ 0.5
          EDV = 0.5 * mc * Vc1 ^ 2 - 0.5 * (mc + mp) * V1 ^ 2

          ' Selección de las Soluciones Factibles -------------------
          If EDV < 0.95 * EDDE Or EDV > 1.05 * EDDE Then
          Else

            ' Incremento en el Contador de Soluciones Factibles -----
            Cont = Cont + 1

            ' Calculo del valor del Funcional ---------------------
              Z = ((fcE - fc) / fcE) ^ 2 + ((fpE - fp) / fpE) ^ 2 + ((dc01E -
              dc01) / dc01E) ^ 2 + ((d12E - d12) / d12E) ^ 2 + ((EDDE -
              EDV) / EDDE) ^ 2

            ' Almacenado de Soluciones Factibles ------------------
            Mat(Cont, 1) = Z
            Mat(Cont, 2) = Vc0
            Mat(Cont, 3) = Vc1
            Mat(Cont, 4) = V1
            Mat(Cont, 5) = EDV
            Mat(Cont, 6) = fc
            Mat(Cont, 7) = fp
            Mat(Cont, 8) = dc01
            Mat(Cont, 9) = d12
            Mat(Cont, 10) = mc
          End If

        Next mc
      Next d12
    Next dc01
  Next fp
Next fc
' FIN DEL PROCESO ITERATIVO *******************************
```

```
'ORDENAMIENTO SEGUN Z (Funcional) ************************
For I = 1 To Cont - 1
   For J = I To Cont
      If Mat(I, 1) > Mat(J, 1) Then
         For K = 1 To 10
            AUX = Mat(I, K)
            Mat(I, K) = Mat(J, K)
            Mat(J, K) = AUX
         Next K
      Else
      End If
   Next J
Next I
'FIN DEL ORDENAMIENTO SEGUN Z *************************

'CREACION Y GRABADO DEL ARCHIVO DE CRITERIO ********
Open "Archivo de Criterio.OUT" For Output As #1

' Impresión de los Datos Principales ------------------------------------
Print #1, "CASO de ESTUDIO"
Print #1, "Masa de la Pick Up [kg] =                          "; mp
Print #1, "Masa del Camion [kg] =                             de 3200 a 8200"
Print #1, "Energia de Deformacion [kNm] =                     ";EDDE/1000
Print #1, "Factor de Friccion Esperado (camion) [ ] =         "; fcE
Print #1, "Factor de Friccion Espedado (pick up) [ ] =        "; fpE
Print #1, "Distancia de Frenado pre-impacto Esperada [m] =    "; dc01E
Print #1, "Distancia de Frenado post-impacto Esperada [m] =   "; d12E

' Impresion de las Soluciones Factibles ---------------------------------
Print #1,"Error  Vc0[km/h]  Vc1[km/h]  V1[km/h]  EDV[kNm]  fc  fp  dc01  d12  mc"

For I = 1 To 60
   contador = contador + 1
   Print #1, Using "######.###"; Mat(I, 1) * 1;
   Print #1, Using "########.#"; Mat(I, 2) * 3.6; Mat(I, 3) * 3.6; Mat(I, 4) * 3.6;
   Print #1, Using "##########"; Mat(I, 5) / 1000;
   Print #1, Using "#######.##"; Mat(I, 6); Mat(I, 7);
   Print #1, Using "########.#"; Mat(I, 8); Mat(I, 9);
   Print #1, Using "##########"; Mat(I, 10)
Next I
Close (1)
'CIERRE DEL ARCHIVO DE CRITERIO ************************

End
```

B.

CONVENTIONAL SOLUTION

VS. PROPOSED METHODOLOGY

B.1 Soludon Conventional

The system of equations to be solved is obviously the same as presented in Section 4.7:

$$V_1 = [2\, g\, d_{12}\, (m_C\, f_C + m_P\, f_P) / (m_C + m_P)]^{0,5}$$
$$V_{C1} = (m_C + m_P)\, V_1 / m_C$$
$$V_{C0} = (V_{C1}^2 + 2\, g\, f_C\, d_{C01})^{0,5}$$
$$E_{D(V)} = 0,5\, m_C\, V_{C1}^2 - 0,5\, (m_C + m_P)\, V_1^2$$

The typical solution is to take the three equations corresponding to V_{e0}; V_{e1} and V1 and assume that the simplest variables (fc / fp / me / mp) and two of the variables obtained from the accident (d_{12} and d_{C01}) are known. As these variables usually have uncertainties (especially in the case of the truck if its state of loading is unknown), they can lead to generate an inappropriate solution. Therefore, although we have three equations and three unknowns (the velocities), the solution can become inconsistent, since when calculating the deformation energy based on the $E_{D(V)}$ velocities determined by the fourth equation, it usually does not coincide with the deformation energy determined based on the $E_{D(D)}$ damage of the vehicles.

The equations of the system to be solved are as follows:

$$\begin{cases} V_1 = [2\ g\ d_{12}\ (m_C\ f_C + m_P\ f_P) / (m_C + m_P)]^{0,5} & \Rightarrow V_1 \\ V_{C1} = (m_C + m_P)\ V_1 / m_C & \Rightarrow V_{C1} \\ V_{C0} = (V_{C1}^2 + 2\ g\ f_C\ d_{C01})^{0,5} & \Rightarrow V_{C0} \end{cases}$$

As it is a system of 3 equations with 3 unknowns, $V1$ V_{C1} and V_{C0} can be determined.

On the basis of V_{Ci} and V_i the Energy that must have been consumed in the deformations during the impact can be calculated:

$$E_{D(V)} = 0,5\ m_C\ V_{C1}^2 - 0,5\ (m_C + m_P)\ V_1^2$$

With the equation corresponding to $E_{D(V)}$ it is possible to determine the eeiergiri that should have been dissipated in the accident, but this energy should be equal to that produced in the deformations of the truck and the van. This equality is usually not achieved and it is common practice to adjust the parameters manually (by hand) to try to achieve a consistent reconstruction.

B.2 Proposed Methodology

In order to avoid this manual work - which in most real cases only allows to obtain an approximate consistency since many variables and equations must be handled simultaneously - this work proposes a mathematical methodology that allows, with scientific rigour, to obtain the best possible results based on the available information.

The proposed solution, to solve these uncertainties, is based on the concept of discrete optimisation, which consists of proposing the same equations that correspond to the mathematical model, but now f_C; f_P; d_{12} and d_{C01} to which mc is added because the amount of truck load is unknown, will be independent or design variables (**X**). These will be varied within a range defined by the expert according to the characteristics of the problem analysed, and with them the dependent variables (**Y**) will be determined.

This methodology requires the inclusion of a functional, which in this problem has been established as an error equation, defined on the basis of the errors that appear between the expected/estimated values according to the expert's criteria and those calculated for each combination of design variables. The feasible solutions will be stored in a file called "criterion file", which will present in an ordered way the obtained results (starting with the combinations that generate the lowest errors).

The expert will analyse the criteria file and take the so-called satisfactory solutions, starting an iterative process that ends when all the results obtained are consistent with all the available information and the expert's criteria.

C.

OTHER TREATMENTS

C.1 Introduction

Parties often wish to know the maximum or minimum speed of one of the vehicles involved in an accident,

A simple strategy for dealing with this type of request is to order the solutions obtained based on the speed of interest, rather than prioritising the minimisation of error. It is important to note that by prioritising a specific variable, a larger margin of error in other variables of the model is accepted.

C.2 Minimum speeds

In order to be able to sort the criteria file according to the velocities from lowest to highest, it is only necessary to make a slight change in the sorting routine. The column number associated with the functional must be replaced by the number corresponding to the column of interest (in the case study, the 1 must be replaced by a 2 in the ***If*** line). This simple change in the index of the matrix will allow to obtain the list of solutions ordered according to the desired speed.

```
'ORDENAMIENTO SEGUN Z (funcional) ************************
For I = 1 To Cont - 1
   For J = I To Cont
    If Mat(I, 1) > Mat(J, 1) Then
    If Mat(I, 2) > Mat(J, 2) Then
     For K = 1 To 10
      AUX = Mat(I, K)
      Mat(I, K) = Mat(J, K)
      Mat(J, K) = AUX
     Next K
    Else
    End If
   Next J
Next I
'FIN DEL ORDENAMIENTO SEGUN Z **************************
```

After making this change you get:

Error	V_{C0}	V_{C1}	V_1	$E_{D(V)}$	f_C	f_P	d_{C01}	d_{12}	m_C
0.227	*90.4*	*72.2*	*45.1*	*242*	*0.40*	*0.80*	*29.0*	*14.5*	*3200*
0.190	90.5	72.4	45.2	244	0.40	0.76	29.0	15.0	3200
0.226	90.6	72.2	45.1	242	0.40	0.80	29.5	14.5	3200

Table C.1: Criterion File - v_{C0} Minimum Velocities

When analysing the file, it is evident that, to reach the minimum speed, the coefficient of friction between the truck and the road would have to be 0.4 (lower limit of the analysed range), while the coefficient of friction of the pick-up would have to be 0.8 (upper limit). The error in these conditions is significantly larger than that of the solution identified in Chapter 4. To obtain more reliable results, the ranges of these variables should be extended to avoid being at their extremes. However, to simplify the analysis, we will assume that the values in Table C.1 will hold even after extending the ranges. Under these conditions, it can be concluded that the low truck-to-road friction coefficient implies very poor adhesion conditions, possibly due to defective brakes or extremely slippery pavement, but since both were driving on the same pavement, a slippery road with a high friction coefficient like that of the pickup is not consistent in principle, so the satisfactory decision in this case might be to check the effectiveness of the truck brakes and make an experimental determination of the friction coefficient in the area of the accident. Once this information is obtained, a new iteration should be performed with this data, continuing the recommended process.

In this particular case, it may not be necessary to run a new cycle after making the above enquiries since even under extreme conditions, the truck speed determined was 90.4 km/h, which is above the maximum limit of 80 km/h. However, the expert should evaluate whether the benefits of obtaining more accurate results justify the cost and time associated with a new iteration involving additional data collection.

C.3 Maximum Speeds

To sort the criteria file by speed from highest to lowest, just make a small adjustment to the sorting routine. It is only necessary to replace the column number indicating the sorting criterion. Instead of using column 1 (functional), column 2 (VC0 speed) will be used. In addition, the comparison sign in the ***If*** condition of the sort loop will be changed from ">" to "<". These simple changes will allow to obtain the desired sorting:

```
'ORDENAMIENTO SEGUN Z (funcional) ************************
For I = 1 To Cont - 1
  For J = I To Cont
  If Mat(I, 1) > Mat(J, 1) Then
  If Mat(I, 2) < Mat(J, 2) Then
  For K = 1 To 10
   AUX = Mat(I, K)
   Mat(I, K) = Mat(J, K)
   Mat(J, K) = AUX
  Next K
  Else
  End If
  Next J
Next I
'FIN DEL ORDENAMIENTO SEGUN Z **************************
```

When running the programme, in which the above-mentioned modifications have been introduced, you get:

Error	V_{C0}	V_{C1}	V_1	$E_{D(V)}$	f_C	f_P	d_{C01}	d_{12}	m_C
0.227	***109.7***	***75.7***	***47.2***	***266***	***0.80***	***0.40***	***31.0***	***13.5***	***3200***
0.173	109.6	75.6	47.1	265	0.80	0.46	31.0	13.0	3200
0.189	109.3	75.1	46.9	262	0.80	0.44	31.0	13.0	3200

Table C.2: Criterion File - Maximum Speeds of v_{C0}

When analysing these data, it is noted that, in order to reach the calculated maximum speed, it is necessary for the friction coefficients to reach again the limits of their ranges, although this time inverted (0.8 for the truck and 0.4 for the pick-up). It can be seen that the error associated with this solution is considerably larger than that obtained in Chapter 4. As in the previous point, the satisfactory solution involves knowing the braking efficiency of the truck and the friction coefficients between the vehicles and the pavement. With this information, the iterative process can be continued.

C.4 Comments

In this section, we have outlined how to proceed when trying to determine minimum or maximum values of a parameter that is not the functional one. Although a specific approach has been proposed, it has not covered all the possible alternatives that the expert could consider depending on his judgement and knowledge of the accident. This approach has been chosen because the main objective of this chapter has been to present the methodology, without going into all the possible details.

Finally, it is important to highlight that the results obtained show that, under all the conditions analysed, and especially in the minimum speed condition, the truck always exceeded the established speed limit of 80 km/h, thus answering the question posed in the Objective defined in Point 4.2.

REFERENCES

[1] Roggero, E. (2024). Efficient Reconstruction of Traffic Accidents. AJEA. Num. 34/2024: 4th Congress on Means of Transport and Associated Technologies. ISBN: 978-950-42-0238-7

[2] Aycock, E. (2015). Accident Reconstruction Fundamentals: A Guide to Understanding Vehicle Collisions. Speakeasy Marketing, Inc. ISBN 13: 9781941645246

[3] Roggero, E. (2010). La Reconstruccion de Accidentes de Transito - Breve Resena Metodologica. Consejo Profesional de Ingenieria Mecanica y Electricista. COPIME - The Journal. ISSN 1668-5857

[4] Roggero, E. (2010). Traffic Accident Reconstruction - Methodology and Example of Application. Professional Council of Mechanical and Electrical Engineering. Proceedings of the event: COPIME - Bicentennial Congresses.

[5] Roggero, E. Cerocchi, M. (2007). Artificial Experience - Application to the Design of Spatial Structures. AATE. Anales del IV Congreso Argentino de Tecnolog^a Espacial.

[6]

Printed by Books on Demand GmbH, Norderstedt / Germany